THE POETRY OF BOHRIUM

The Poetry of Bohrium

Walter the Educator

Silent King Books

SILENT KING BOOKS

SKB

Copyright © 2024 by Walter the Educator

All rights reserved. No part of this book may be reproduced in any manner whatsoever without written permission except in the case of brief quotations embodied in critical articles and reviews.

First Printing, 2024

Disclaimer
This book is a literary work; poems are not about specific persons, locations, situations, and/or circumstances unless mentioned in a historical context. This book is for entertainment and informational purposes only. The author and publisher offer this information without warranties expressed or implied. No matter the grounds, neither the author nor the publisher will be accountable for any losses, injuries, or other damages caused by the reader's use of this book. The use of this book acknowledges an understanding and acceptance of this disclaimer.

"Earning a degree in chemistry changed my life!"
- Walter the Educator

dedicated to all the chemistry lovers, like myself, across the world

BOHRIUM

In the cosmic dance of elemental flair,

BOHRIUM

Where atoms fuse in the celestial air,

BOHRIUM

Lies Bohrium, a wondrous element rare,

BOHRIUM

With mysteries concealed beyond compare.

BOHRIUM

Born from nuclear fusion's fiery embrace,

BOHRIUM

In stars' cauldron, it finds its place,

BOHRIUM

A fleeting visitor in cosmic space,

BOHRIUM

Yet in our world, a fleeting trace.

BOHRIUM

Unseen by mortal eyes, it roams,

BOHRIUM

In the depths of atomic homes,

BOHRIUM

A silent sentinel, it silently combs,

BOHRIUM

The universe's vast, cosmic tomes.

BOHRIUM

In labs of science, where minds aspire,

BOHRIUM

Bohrium reveals its atomic desire,

BOHRIUM

A synthetic creation, forged in fire,

BOHRIUM

By human hands, it rises higher.

BOHRIUM

With eighty-seven protons, it stands,

BOHRIUM

In the periodic table's grand bands,

BOHRIUM

An enigma shrouded in scientific strands,

BOHRIUM

In laboratories, where intellect expands.

BOHRIUM

Yet beyond its atomic shell,

BOHRIUM

Bohrium's tale begins to swell,

BOHRIUM

A story of discovery, where scientists dwell,

BOHRIUM

In quests for knowledge, they do excel.

BOHRIUM

In the heart of the atom, where chaos reigns,

BOHRIUM

Bohrium's essence, the mind constrains,

BOHRIUM

A symbol of persistence that science gains,

BOHRIUM

In the pursuit of truth, where no one feigns.

BOHRIUM

Its fleeting existence, a testament bold,

BOHRIUM

To the wonders of science, to behold,

BOHRIUM

In laboratories, where stories unfold,

BOHRIUM

Bohrium's secrets, forever told.

BOHRIUM

From the depths of the universe's embrace,

BOHRIUM

To the confines of Earth's humble space,

BOHRIUM

Bohrium whispers tales of cosmic grace,

BOHRIUM

In every atom, in every place.

BOHRIUM

So let us marvel at Bohrium's lore,

BOHRIUM

And delve into its mysteries, explore,

BOHRIUM

For in its essence, we find much more,

BOHRIUM

Than just an element, but a cosmic door.

BOHRIUM

To realms unseen, where knowledge gleams,

BOHRIUM

In the vastness of scientific dreams,

BOHRIUM

Bohrium beckons with celestial beams,

BOHRIUM

A beacon of discovery, it seems.

BOHRIUM

ABOUT THE CREATOR

Walter the Educator is one of the pseudonyms for Walter Anderson. Formally educated in Chemistry, Business, and Education, he is an educator, an author, a diverse entrepreneur, and he is the son of a disabled war veteran. "Walter the Educator" shares his time between educating and creating. He holds interests and owns several creative projects that entertain, enlighten, enhance, and educate, hoping to inspire and motivate you.

Follow, find new works, and stay up to date
with Walter the Educator™
at WaltertheEducator.com

www.ingramcontent.com/pod-product-compliance
Lightning Source LLC
LaVergne TN
LVHW052005060526
838201LV00059B/3846